Table of Contents

List of Figures

List of Tables

1 Introduction to Financing Solar Installations on K–12 Public Schools

Solar energy systems installed on public schools have a number of benefits that include utility bill savings, reductions in greenhouse gas emissions (GHGs) and other toxic air contaminants, job creation, demonstrating environmental leadership, and creating learning opportunities for students. In the 2011 economic environment, the ability to generate general-fund savings as a result of reducing utility bills has become a primary motivator for school districts trying to cut costs. To achieve meaningful savings, the size of the photovoltaic (PV) systems installed (both individually on any one school and collectively across a district) becomes much more important; larger systems are required to have a material impact on savings. Larger PV systems require a significant financial commitment and financing therefore becomes a critical element in the transaction.

In simple terms, school districts can use two primary types of ownership models to obtain solar installations and cost savings across a school district. The PV installations can be financed and owned directly by the districts themselves. Alternatively, there are financing structures whereby another entity, such as a solar developer or its investors, actually own and operate the PV systems on behalf of the school district. This is commonly referred to as the "third-party ownership model." Both methods have advantages and disadvantages that should be weighed carefully.

1.1 Direct Ownership

If a district owns its PV systems, then it receives all of the electricity savings and any available rebates, and retains the associated renewable energy certificates (RECs) that allow a district to make environmental claims about its PV systems. Thus, savings to the general fund that result from reduced or eliminated utility bills can be used to repay the loan or bond that was used to purchase a PV system. The district can also publicly claim to be reducing its GHG and toxic air contaminant emissions.

When a school district uses voter-approved general obligation bonds for the purchase of a PV system, or when the cost of repaying the debt incurred in purchasing the PV system is less than the utility savings, these excess funds can be used for other needed school services. Also, to compensate for the inability to directly benefit from federal tax incentives, the California Solar Initiative (CSI) incentives are greater for governmental and nonprofit organizations than for commercial entities. [1] The CSI incentives, when available, provide another revenue stream for a school district. Finally, given the productive life of a PV system (25–40 years), it is likely that any debt incurred to finance the PV will be paid off well before the end of the useful life of the system.

A primary disadvantage of direct ownership is the capital commitment involved. School districts rarely have cash reserves and might not have voter-approved bonding authority or access to the financial mechanisms needed to purchase PV systems. The district also simply might be unwilling to incur any new debt. Additionally, with a direct purchase, the school district is responsible for operations and maintenance (O&M) for the systems, unless it signs a long-term maintenance agreement with the solar developer—an option that is becoming increasingly common. Lastly, the federal tax credits available to tax-paying entities are not available for public entities directly purchasing a PV system.

1.2 Third-Party Ownership

The advantages of third-party financed PV installations for school districts include: little to no capital investment is required on the part of the school district; districts are not responsible for O&M; private-sector tax incentives can be incorporated into the transaction, which should result in reduced cost of the electricity sold to the district; and the districts can purchase the system for "fair market value" (FMV) during or at the end of the contracted term.

The disadvantages of the various third-party finance models, and of the third-party power purchase agreement (PPA) model in particular, include the following:

- A PPA is a complicated transaction that requires the school district to invest time and money in assuring that it negotiates a fair and equitable contract. Utility bill savings will be less than if the districts directly owned the system, because 100% of the solar electricity generated by the PV system must be purchased by the schools from the third-party investor.

- A PPA generally allocates the RECs to the investor, in which case the district is not entitled to claim the environmental benefits associated with clean electricity production.

- If, in the future, the district decides to purchase the PV system from the investor, there is no way to determine the purchase price in advance because the system must be sold for its fair market value at the time of sale.

- Unless the district exercises its buyout option to purchase the system at the end of the PPA term, the school will not own the PV system. In such cases, the PPA should include a provision for the removal of the system at the investor's expense.

Regardless of how a school district decides to finance or acquire PV installations on its buildings, several key issues should be highlighted, including benchmarking of district energy use, energy efficiency, PV siting considerations (including roof condition), the potential for school closings during project installation, and the integration of the PV system into the classroom environment.

1.3 Energy Efficiency and Benchmarking

Energy efficiency improvements if not already undertaken should be incorporated in the planning stages prior to installing a PV system. Energy conservation measures (ECMs) are the most cost-effective way to save energy and realize utility-bill savings. The return on investment in PV can be enhanced when the building hosting the system is already energy efficient. Furthermore, to qualify for rebates under the California Solar Initiative, an energy audit of existing buildings is required. For new schools, the project is required to meet either of the following two tiers of energy efficiency.

- Tier I—15% reduction in the commercial building's combined space heating, space cooling, lighting, and water-heating energy compared to the 2008 Title 24 Standards.

- Tier II—30% reduction in the commercial building's combined space heating, space cooling, lighting, and water-heating energy compared to the 2008 Title 24 Standards.

Achieving the Tier I level is the minimum condition required to qualify for the rebate. Tier II is the preferred level that builders are encouraged to meet. For either Tier I or II, any equipment or appliance provided by the builder must be ENERGY STAR-labeled if this designation is applicable. [2] Schools can implement energy efficiency measures either prior to the PV installations or in combination with them. An energy savings performance contract (ESPC) is a mechanism to finance the energy efficiency upgrades and possibly the PV installations as well. These contracts are discussed elsewhere in this document.

To identify schools that are most in need of energy efficiency upgrades, the schools' energy use should be benchmarked. A simple way to do this is to compare the energy intensity—the energy use per square foot, determined by dividing total annual energy use by total facility square footage—for all schools. Those with the greatest energy intensity should be a priority for energy audits and ECM identification. The U.S. Environmental Protection Agency (EPA) offers a free benchmarking tool called Portfolio Manager, which can help identify the energy performance of a district's schools. High-performing schools are eligible for an ENERGY STAR certification.[1] In some cases, the local utility can assist with energy audits and might provide rebates for many of the audit's recommended upgrades.

1.4 Solar Photovoltaics

Photovoltaic arrays convert sunlight to electricity without moving parts and without producing fuel wastes, air pollution, or GHGs. They require very little maintenance and make no noise. Arrays can be mounted on all types of buildings and structures, as well as in parking lots or other open spaces. A PV system's direct current (DC) output can be conditioned into grid-quality alternating current (AC) electricity, or DC can be used to charge storage batteries. Most systems installed on schools do not generally have batteries because they are cost-prohibitive.

Traditional solar cells are made from silicon and are usually flat-plate. Second-generation solar cells are called thin-film solar cells because they are made from amorphous silicon or non-silicon materials such as cadmium telluride. Thin-film solar cells use layers of semiconductor materials only a few micrometers thick. Because of their flexibility, thin-film solar cells can double as rooftop shingles and tiles, building facades, or the glazing for skylights.

The cost of PV-generated electricity has dropped nearly twentyfold in the last 40 years. Grid-connected PV systems currently sell for about $5 to $8 per watt-peak (Wp) ($0.20 to $0.30 per kilowatt-hour), including support structures and power conditioning equipment. An NREL study of 7,074 PV systems installed in 2007 reported a range of total capital cost averaging from $8.32 per watt (W) for small systems (less than 10 kW) and $6.87 per watt for large systems (greater than 100 kW). In April 2011, West Contra Costa Unified School District received bids for an approximately 310-kW ground-mounted system that cost about $6.39 per watt (DC) including a 10-year performance guarantee and a 10-year O&M contract. Operations and maintenance costs are reported at $0.008 per kilowatt-hour produced, or at 0.17% of capital cost without tracking

---

[1] Portfolio Manager is available at
http://www.energystar.gov/index.cfm?c=evaluate_performance.bus_portfoliomanager. Accessed June 8, 2011.

and 0.35% with tracking. [3] Traditional silicon PV panels are very reliable and last 25 years or longer; most panels come with 20- to 25-year warranties. [4]

1.4.1 Siting Photovoltaics

The major challenge of siting solar PV technologies is determining the appropriate siting for maximum electricity production. An ideal solar installation is situated in an unshaded, south-facing location with an optimum tilt angle and supplies electricity to a site where there is adequate demand for the electricity produced.[2]

Not all sites are suitable for solar technologies. There are a few important rules of thumb that might be helpful in determining whether solar technologies are appropriate for a site.

- It is important to identify an area that is unshaded, particularly during the peak daylight hours of 10 a.m. to 2 p.m. Shade not only reduces the output of the solar panel being shaded, but it also can reduce the output of adjacent panels, even if they are still in the sun. Shade can be caused by trees, nearby buildings, and roof equipment or features (such as chimneys).

- It generally is best to orient fixed-mount panels due south in the Northern Hemisphere. Siting panels so that they face east or west of due south decreases electrical output; that effect varies by location, however, and at some locations the loss of efficiency could be minimal.

- The optimal tilt angle for achieving the best performance from a fixed-mount PV panel is a tilt angle equal to the latitude of a location, for locations in latitudes less than 20 degrees. At higher latitudes the correlation is not valid. Christensen and Barker analyzed the annual solar resource data for different latitudes. [5] To maximize the annual energy production at a location of 40° north latitude, the optimal tilt varies from 30° to 35°.

- Fixed-mount solar panels can be flush- or tilt-mounted on roofs, be pole-mounted on the ground (e.g., a carport structure), or be integrated into building materials such as roofs, windows, and awnings. A tilt angle equal to latitude is not always feasible, however, perhaps because of roof pitch or wind or snow loading considerations. It is possible to install panels at a different angle. The effect of tilt angle varies by location and in some locations could be minimal.

- The size and nature of a school's electric load must be well understood to properly select and size a PV system. Photovoltaic systems can be designed to power any electrical load regardless of size or location, as long as sunlight and space for the panels are available. Likewise, the systems can be designed to power any percent of an electric load, from a very small percentage to more than 100% of the load, depending on the area available for the panels and the availability of the sun. When considering a system that will be tied to the utility grid (grid-connected), it is essential

---

[2] Efforts are underway in the California state legislature and CPUC to allow PV systems to be built on sites that have minimal loads and to allow the value of the electricity produced to be taken as a bill credit at another less solar-friendly site.

to understand the applicable net metering rules and interconnection standards for the local electric utility company.

- For electric customers that generate their own electricity, net metering—when allowed by the serving utility—enables the customer to earn a bill credit if and when the on-site solar PV system generates more electricity than is used on site. This typically is accomplished by use of a single, bidirectional meter. When a customer's generation exceeds its use, electricity from the customer flows back to the grid, offsetting electricity consumed by the customer at a different time during the same billing cycle. In effect, the customer uses excess generation to offset electricity that the customer would otherwise have to purchase at the utility's full retail rate. Net metering is required by law in most states, but these policies vary widely. [6]

- Interconnection standards specify the technical and procedural process by which a customer connects a PV system that generates electricity to the grid. Such standards include the technical and contractual arrangements by which system owners and utilities must abide. State public utilities commissions typically establish standards for interconnection to the distribution grid, but the Federal Energy Regulatory Commission (FERC) has adopted standards for interconnection at the transmission level. Many states have adopted interconnection standards, but some states' standards apply only to investor-owned utilities (IOUs)—not to municipal utilities or electric cooperatives. Several states have adopted interconnection "guidelines," which are weaker than standards and generally apply only to net-metered systems. [6]

- Photovoltaic modules have different efficiencies (i.e., higher efficiency panels produce more electricity per unit area than lower efficiency panels), therefore it is important to consider the efficiency versus the available or required area of the PV system. Fewer modules made of a higher-efficiency cell (such as single-crystalline) are needed for approximately the same power output as produced by more modules that are made of a lower-efficiency cell (such as thin film). Therefore, if a project location is space-constrained, then a module that is more efficient—and potentially costs more—might make the most sense. If a project has an abundance of space, however, then a less-efficient, less-costly module could be the most cost effective.

1.4.2 Roof Condition

Photovoltaic systems should only be installed on roofs that are in good shape and which can reasonably be expected to remain in good condition for the entire expected lifetime of the PV system (at least 25 years). Roofs should therefore be relatively new or be upgraded prior to the PV installation. It generally is not cost effective to remove a previously installed PV system to replace or upgrade a roof, although certain rooftop-PV mounting systems now make it possible to upgrade a roof without removing the PV structure. Ideally, roofs that need repair or are slated for a replacement can be improved or replaced in conjunction with the installation of the PV system. Structural assessments might also be required to confirm that the roof can support the additional weight and wind loading. If the best sites for solar are those that need new or improved roofs, then this near-term capital expense must be budgeted for accordingly. Building-integrated PV systems that combine the roofing material with the PV installation (thin-film applications) could also be an option.

1.4.3 School Closings

With an expected life of 25 years or more, once installed, a PV system will generate electricity to offset the building's load for a long time. To the degree possible, it is best to identify sites that are expected to remain in service for the foreseeable future. Given the current economic climate and the stress on school budgets, school closings are unavoidable. Although a PV system can be removed from a roof and reinstalled elsewhere in the district, this can be a costly process that also results in lost electricity production during system downtime. State regulations are evolving on the issue of whether the electricity generated by a PV system has to be consumed on site. It might be possible in the future to continue generating electricity from a school that has been closed and to apply the value of the generated electricity to another electricity account in the district.

1.4.4 Classroom Impact

Utility-bill savings are becoming the primary motivation for school districts to install PV systems. The impact of an on-site solar installation, however, goes beyond the value of the electricity produced and the greenhouse gases avoided. Photovoltaic installations sited throughout a school district create an excellent platform to introduce energy issues to students, teachers, and the school community, and provide hands-on experience for an issue that is traditionally given little attention in standard K–12 curriculums. Therefore, incorporating the PV systems into all school curricula should be a key element in any district-wide solar program. To maximize the classroom impact, the following are key activities:

- Curriculum development

- Data acquisition/monitoring system with Web access

- Training for facility staff

- Training for teachers

- Kiosks or other appropriate signage

On large system purchases, some of these activities could be provided by the PV provider as part of the negotiated contract. With these general concepts in mind, the remainder of this introductory report focuses on the financial alternatives available to school districts as they implement their Solar Master Plans.

2 Direct Ownership of Photovoltaic Systems

In fiscal year (FY) 2009–2010, local governments and school districts were often the beneficiaries of low-interest or 0% interest bonds backed by the federal government. In some cases the bonds have been used to purchase PV systems. Prior to 2009, school districts had to think creatively if they wanted their schools to become energy generators. Some California school districts chose to enter into power purchase agreements. Other districts used voter-approved bonds, school modernization grants from the state, and up-front rebate payouts to help underwrite the cost of their PV systems.

It doesn't seem likely that the federal government will reauthorize the bond programs that helped to build so many solar projects. Additionally, rebates in California are evaporating and might not be replenished. On the plus side, solar-panel costs have been dropping, their efficiency levels have been increasing, and the slowed economy has made the cost for construction projects much more competitive and favorable for solar projects. An emerging market for a Tradable Renewable Energy Credit (TREC) in California could provide some additional financial incentives lost when the CSI rebates end. The California Energy Commission (CEC) has the authority to allow school districts to participate in the TREC market but, as of May 2011, the question of whether the CEC will use this authority remains unanswered.

One way to avoid the "boom-and-bust" cycle associated with PV financing for school districts is to incorporate the cost of solar installations into the next request that a district makes to its residents for general obligation bonds that support school construction projects. General obligation, tax-exempt, municipal bonds are common financing tools for schools. Photovoltaic projects can be bundled with other investments into a much larger bond transaction. The bond cycle is relatively infrequent for school districts (every 5 to 10 years), so planning is critical if these bonds are going to be used for PV installations. [7] The pursuit of Solar Master Plans is a key element in this planning process. Specific sites can be identified, their solar resources can be characterized, and an estimate of costs can be determined to create a priority list of installations. By creating this list of qualified projects a district will be ready to include them in the next funding cycle, instead of inserting vague language stating that some of the proceeds will be used for renewable energy projects and taking the risk of losing them to other investment priorities.

A school district can directly purchase, own, and operate PV systems using a variety of financing mechanisms. These include using existing reserves available from the General Fund, traditional tax-exempt bond financing, proceeds from state transfers of funds (e.g., state school construction and modernization funds) and other forms of grants (e.g., from foundations and private businesses), and a variety of tax credit bonds. With the exception of tax credit bonds, the other mechanisms are relatively common ways that school districts traditionally finance their capital investments and are not discussed in detail. Utility rebates, if available, also can be used to supplement the financing of the PV system.

As noted, CSI incentives are greater for school districts under the direct-ownership scenario. Check the CSI Statewide Trigger Point Tracker regularly for the status of rebates from California's major investor-owned utilities. [1] As of May 4th, 2011, CSI incentives were suspended in Pacific Gas and Electric Company (PG&E) and San Diego Gas and Electric (SDG&E) territory for all sectors except residential installations. [1] Legislation (S.B. 585 Kehoe) is attempting to replenish the CSI so that it can fulfill its original legislative mandate. Check the Database of State Incentives for Renewables and Efficiency (DSIRE) for all state and federal incentives, including rebates from publicly owned utilities.[3] Municipal utilities or publicly-owned utilities (POUs) also offer solar rebates; however, the cost of the electricity delivered by a municipal utility is sometimes too low to make districts served by POUs attractive candidates for PPAs. If a school district decides to finance and own a solar energy system, it can certainly finance it with voter-approved general obligation bond proceeds and other forms of traditional tax-exempt financing, or it could possibly use cash on hand if available.

---

[3] *See* http://www.dsireusa.org/. Accessed June 8, 2011.

2.1 Using Cash on Hand

Although it is unlikely in the current economic environment that a school district has available general-fund resources on hand to directly purchase a PV system without financing, it is not out of the question. A district could be the recipient of grant funding or, as a result of a sale of unused property, could have the resources to purchase and install a PV system. If this is the case, then the school district would install the system and immediately begin accruing utility bill savings. The CSI production incentives, if available when the project is initiated, would enhance this positive cash flow in the first 5 years. Simple calculators can be developed to illustrate these savings to a school district.

2.2 California Energy Commission Loans for Energy Efficiency and Renewable Energy

Using funds from a variety of sources, including federal stimulus dollars, the CEC has a low-interest loan program available for public entities, including schools. [8] The list of eligible projects includes renewable energy in addition to a host of energy efficiency measures. The interest rate of the loans is 3% per annum and the maximum term cannot exceed 15 years or the expected life of the equipment (whichever is less). For PV systems, 15 years is less than the expected system life, thus 15 years would be the maximum term. The loan is repaid using the energy savings. Loans are given on a first-come, first-served basis and are based on available funding. For more information, consult the CEC website at http://www.energy.ca.gov/efficiency/financing/index.html.

2.3 Other Tax-Exempt Financing

2.3.1 Tax-Exempt Municipal Leasing

Leasing equipment instead of purchasing it is a common way for schools to finance certain hard assets (e.g., vehicles, software, computers, office equipment). Leasing is used much less frequently, however, for solar installations. This is a function of the inability of the owner of the PV system (the "lessor") to receive the federal tax incentives, given that the school, as the user of the equipment (the "lessee"), is not subject to U.S. income taxes. Investment tax credits are so valuable that alternatives to a tax-exempt lease often are more attractive. For some school districts, however, the low cost and familiarity of a tax-exempt lease combined with greater incentives of the state rebate program and the ability to execute a lease without voter approval could outweigh the loss of the tax credits in the transaction. Information on Yolo County, California, which used a tax-exempt lease as part of the capital structure to finance 1 MW of PV energy, can be found at http://www.nrel.gov/docs/fy11osti/49450.pdf.

In early 2010, another option for leasing was created under the U.S. Treasury's 30% Cash Grant in Lieu of the Investment Tax Credit program (the "1603 Program"). [9] A third party who elects to receive the cash grant to finance a PV system instead of taking the 30% Investment Tax Credit (ITC) can lease this system to a school despite its tax-exempt status. [10] Although certain caveats are associated with this structure—such as the inability to benefit from accelerated depreciation—it does create an additional option for schools to consider. [10] The U.S. Treasury cash grant program was set to expire at the end of 2010 but has been extended by one year. Although the authors are unaware of any use of this mechanism to lease PV systems to schools, it does remain an alternative for school districts to consider through the end of 2011.

2.3.2 Office of Public School Construction Funds

A potential source of funds for solar projects could be the State of California's Office of Public School Construction (OPSC) and the High Performance Incentive (HPI) Program. The OPSC implements and administers the School Facility Program (SFP), which includes the New Construction Grants and Modernization Grants, and other programs of the State Allocation Board (SAB). The HPI Program was established to distribute funds set aside for high energy performing schools to promote the use of high-performance attributes in new construction and modernization projects for K–12 schools. The HPI awards credits through a scorecard tied to the 2006 Collaborative for High Performance Schools (CHPS) guidelines, which determine the HPI points and the HPI amount that the school can receive.

2.3.2.1 New Construction Grant

The New Construction Grant offered by OPSC provides state funds on a 50/50 state-local sharing basis for public schools' capital facility projects in accordance with the statute. Eligibility for state funding is based on a district's need to house pupils and is determined by criteria set in law. This new construction grant amount is intended to provide the state's share for all necessary project costs except those for site acquisition, utilities, and off-site, service-site, and general-site development that might qualify for additional project funding. The necessary project costs include, but are not limited to, funding for design and the construction of the building, educational technology, tests, inspections, and furniture/equipment.

2.3.2.2 Modernization Grant

The modernization grant made by OPSC provides state funds on a 60/40 basis for improvements to educationally enhance school facilities. Projects eligible under this program include upgrades to air conditioning systems, plumbing, lighting, roof replacement, PV systems, and electrical systems. Site acquisition cannot be included in modernization applications. The modernization grant amount is intended to provide the state's share for all necessary project costs. The necessary project costs include, but are not limited to, funding for design and the modernization of the building, educational technology, tests, inspections, and furniture/equipment. School districts typically use local bond financing or secure alternative funding to meet the 50% funding requirement for new construction projects or the 40% funding requirement for modernization projects. The application filing timelines are presented in Table 1.

Table 1. Application Filing Timelines

Program[a] / Type of Application	Application Acceptance Date	Application Due Date
New Construction		
Design [b]	Ongoing	Prior to occupancy of any of the classrooms
Separate Site [a,c]	Ongoing	Prior to occupancy of any of the classrooms
Construction (Full Adjusted Grant)	Ongoing	Prior to occupancy of any of the classrooms
Modernization		
Design [a,c]	Ongoing	None[c]
Construction (Full Adjusted Grant)	Ongoing	None[a,c]

a. For application submission requirements, see the OPSC website, http://www.dgs.ca.gov/Default.aspx?alias=www.dgs.ca.gov/opsc, and the SFP Regulations, http://www.bondaccountability.opsc.dgs.ca.gov/bondac/oversight_K12.asp.
b. Application only can be submitted if the district qualifies for financial hardship assistance.
c. Applications accepted for reimbursement for any contracts signed after August 27, 1998.

Table 2 presents the status of the funds. Proposition 1D, Proposition 47, and Proposition 55 have an available combined balance for new construction and modernization of $3 billion as of May 25, 2011. Most of this amount, however, appears dedicated to activities other than energy investments. The School Facility Program requirements for the New Construction Grant and Modernization Grant can be found at www.documents.dgs.ca.gov/opsc/Resources/SFP_NC_Rqmnts.pdf and www.documents.dgs.ca.gov/opsc/Resources/SFP_Mod_Rqmnts.pdf.

Table 2. Status of Funds

STATUS OF FUNDS				
Per June 22, 2011 State Allocation Board Meeting				
(Amounts In Millions of Dollars)				
Program	Remaining Bond Authority As of May 25, 2011	Apportionments - June 22, 2011	Adjustments After Board - June 22, 2011	Remaining Bond Authority As of June 22, 2011
Proposition 1D				
New Construction	$32.5			$32.5
Seismic Repair	194.8			194.8
Modernization	1,350.5			1,350.5
Career Technical Education	120.1	3.4		123.5
High Performance Schools	79.0			79.0
Overcrowding Relief	509.2	4.8		514.0
Charter School	409.3			409.3
Joint Use	0.6			0.6
Sub-total	$2,696.0	$8.2	$0.0	$2,704.2
Proposition 55				
New Construction	$721.5			$721.5
Modernization	5.8			5.8
Critically Overcrowded Schools				
Reserve	286.7			286.7
Charter School	132.2			132.2
Relocation/Dtsc Fees	13.1			13.1
Hazardous Material/Waste Removal	2.6			2.6
Conversion Increase Fund	23.3			23.3
Joint Use				
Sub-total	$1,185.2	$0.0	$0.0	$1,185.2
Proposition 47				
New Construction	$109.9			$109.9
Energy	0.1			0.1
Modernization	4.3	0.2		4.5
Critically Overcrowded Schools				
Reserve				
Charter School	45.9			45.9
Conversion Increase Fund	15.6			15.6
Joint Use				
Sub-total	$175.8	$0.2	$0.0	$176.0
Sub-total	$4,057.0	$8.4	$0.0	$4,065.4
Proposition 1A				
New Construction	$0.9			$0.9
Modernization	0.6			0.6
Hardship	6.2	0.3		6.5
Class Size Reduction				
Sub-total	$7.7	$0.3	$0.0	$8.0
SFP Total	$4,064.7	$8.7	$0.0	$4,073.4
Williams Lawsuit				
Needs Assessment Program				
Emergency Repair Program	$457.1 ^			457.1
Sub-total	$457.1	$0.0	$0.0	$457.1
GRAND TOTAL	$4,521.8	$8.7	$0.0	$4,530.5

Note:

A. Funds are not available at this time.

Source: http://www.documents.dgs.ca.gov/opsc/Resources/Funds_Status.pdf

2.3.2.3 High Performance Incentive Program

In 2006, the HPI program was established to distribute the $100 million set aside for high-performance schools from Proposition 1D to promote the use of high-performance attributes in new construction and modernization projects for K–12 schools. On the Status of Funds (dated January 26, 2011), the High Performance Schools Program had an available balance of $80.5 million. [11] The School Facility Program regulations were based on 2006 California Collaborative for High Performance Schools (CA-CHPS) and referenced the 2005 Title 24 standards. According to the Division of the State Architect (DSA) website, the 2009 CA-CHPS Criteria now are accepted for the DSA/HPI grant review.[4] The HPI points are calculated from a project scorecard. The HPI project scorecard was based on the 2006 CHPS guidelines, which remain unchanged. The HPI amount is based on the points attained by the district within the following five categories: Site, Water, Energy, Materials, and Indoor Environmental Quality. The DSA's High Performance Section (HPS) verifies the HPI rating criteria to determine the number of points the project receives. A checklist for HPI projects and the DSA/HPI scorecards/guidelines can be found on these websites, respectively: www.documents.dgs.ca.gov/dsa/other/HPI_Checklist_rev02-07-10.pdf and www.dgs.ca.gov/dgs/tabid/1378/Default.aspx#t4 .

Table 3. High Performance Incentive Points Summary [12]

Modernizations and Additions	
Minimum to Qualify	20
Maximum	77
New Construction (New Campus Only)	
Minimum to Qualify	27
Maximum	75

2.4   Qualified Tax Credit Bonds

A number of qualified tax credit bonds (QTCB) have proven to be suitable vehicles for financing solar installations on schools, including Clean Renewable Energy Bonds (CREB), Qualified Energy Conservation Bonds (QECB), Qualified School Construction Bonds (QSCB), and Qualified Zone Academy Bonds (QZABs). Unfortunately, QTCBs are no longer available but are included here for informational purposes. Some school districts may still have access to prior years' allocations, and it is possible that some form of QTCB could be made available in the future.

By providing allocations of federal tax credits for certain categories of projects, the cost of capital is reduced and, ideally, more of these projects are built. The CREBs and QECBs are tax credit bonds aimed at renewable energy and energy efficiency investments. The QSCBs and QZABs are directed at schools and are defined broadly enough to also include renewable energy and energy efficiency.

2.4.1 Build America Bonds

Although not technically a QTCB, Build America Bonds (BABs) have a tax credit feature similar to that of the QTCBs. Their success, however, has been a result of what is known as the

---

[4] *See* www.dsa.dgs.ca.gov (accessed June 9, 2011).

"direct payment" option. Instead of BAB buyers receiving federal tax credits in lieu of interest payments, the issuer can elect to receive a subsidy from the U.S. Treasury. This subsidy is equivalent to 35% of the bond's interest rate. Therefore, it is possible for state and local governments, including school districts, to issue taxable bonds that actually are cheaper than tax-exempt bonds once the subsidy is included. As a result, BABs have been very successful since the program's creation in 2009. To date, more than $120 billion in BABs have been issued. [13]

According to the *Bond Buyer*, although initial BAB transactions were large (for example, the first was a $250-million bond issued by the University of Virginia), the average size of a BAB issuance is decreasing; bonds in the $1-million to $5-million range now are more common. A bond of this amount could be issued as a dedicated solar bond for an individual school district. Note that, as of March 2011, there was no reauthorization of funding for BABs. Funding could possibly occur in late 2011.

2.4.2 *Hiring Incentives to Restore Employment Act of 2010 and the Impacts on Qualified Tax Credit Bonds*

The Hiring Incentives to Restore Employment (HIRE) Act of March 2010 made a very significant modification to the CREB, QECB, QSCB, and QZAB tax credit bond programs, creating a direct-pay subsidy mechanism similar to the BAB program (but much more generous). Under the new HIRE Act provisions, the subsidy that the issuer of a direct-pay bond receives is the lesser of either the actual interest rate of the bond or the reference credit rate found on the Treasury Direct website.[5] As an example, on December 14th, 2010, the reference credit rate on the Treasury Direct website was 5.63% (annual rate) for a qualified tax credit bond with a maximum maturity of 18 years. The QSCB and QZAB issuers get a direct-pay subsidy equal to 100% of the applicable tax credit rate of 5.63%. The CREBs and QECBs receive 70% of the applicable rate, which is 3.94%. Therefore:

- If a QSCB or QZAB was issued on December 14th, 2010, then the interest rate subsidy the issuer receives is the lesser of the actual interest rate of the bond or 5.63%. In other words, any bond with an interest rate of 5.63% or less is, in effect, an interest-free bond because the government subsidy offsets the entire interest payment. If the interest rate is more than 5.63%, then the net interest cost to the issuer is the difference between the actual rate and 5.63%.

- If a CREB or QECB was issued on December 14th, 2010, then the interest rate subsidy the issuer receives is the lesser of the actual interest rate of the bond or 3.94%. In other words, any bond with an interest rate of 3.94% or less is, in effect, an interest-free bond because the government subsidy offsets the entire interest payment. If the interest rate is more than 3.94%, then the net interest cost to the issuer is the difference between the actual rate and 3.94%.

For school districts with access to allocations of different types of tax credit bonds, issuing QSCBs or QZABs is more likely to result in interest-free financing, given the greater subsidy available for these bonds versus CREBs and QECBs.

---

[5] *See* "Qualified Tax Credit Bond Rates." *TreasuryDirect.* https://www.treasurydirect.gov/GA-SL/SLGS/selectQTCDate.htm. Accessed June 9, 2011.

2.4.3 Clean Renewable Energy Bonds

Initially authorized under the Energy Policy Act (EPAct) of 2005, Clean Renewable Energy Bonds (CREBs) [14] are an attempt to level the playing field for public entities unable to benefit from the tax incentives available to private entities. These bonds must be used for qualified renewable energy projects, which include PV. In 2009, the State of California received $640 million—80% of the total amount allocated for local governments in the United States. Most of the California allocations are for solar projects. Many California school districts received CREB allocations, including the Oakland Unified School District, which received 17 separate allocations for a total of $39 million. Berkeley and West Contra Costa school districts do not appear on the IRS list as having received any CREB allocations.

Clean Renewable Energy Bonds can also be combined with other tax credit bonds or with more traditional tax-exempt financing. In early 2010, for example, Yolo County, California, combined CREBs, QECBs, a California Energy Commission Loan, and a tax-exempt municipal lease to finance a 1-MW solar installation on the Yolo County Justice Center. [15] Additionally, the project is receiving a CSI incentive of $0.24/kWh for 5 years. Bank of America Corporation structured this transaction. This project is noteworthy in that it is one of the first QECB issuances in the country and was the first to combine QECBs with CREBs. [15] A total of $7.265 million was raised across the four financial products. This transaction was completed prior to the HIRE Act coming into effect; therefore, the CREBs and QECBs are using the tax credit feature in which the buyer receives a federal tax credit in addition to a 3.90% supplemental interest payment from Yolo County. More information on this transaction can be found at http://www.nrel.gov/docs/fy11osti/49450.pdf. Note that, as of March 2011, there was no reauthorization of funding for CREBs.

2.4.4 Qualified Energy Conservation Bonds

A Qualified Energy Conservation Bond (QECB) is very similar to a CREB. Unlike CREBs, however, up to 30% of QECBs can be used to finance private-sector activity. Also, there are numerous additional renewable energy and energy conservation projects that can be financed with QECBs, one of which is capital expenditures for reducing energy consumption in publicly owned buildings by at least 20%. [16] This is relevant for those cases in which a school district plans to finance energy efficiency upgrades in addition to installing PV systems.

Unlike CREBs, which required submitting an application to the IRS to solicit a tax credit allocation, the QECB tax credits were allocated to states based on population. This state-by-state allocation occurred in 2009. California received an allocation of approximately $381 million. [16] Cities and counties in California that have populations greater than 100,000 automatically received sub-allocations of this amount, with $170 million going to cities, $198 million to counties, and the remaining $13 million to state and tribal governments. [17] According to the California Debt Limit Allocation Committee (CDLAC), the cities of Oakland, Berkeley, and Richmond received QECB allocations of approximately $4 million, $1 million, and $1 million, respectively. [18] Note that, as of March 2011, there was no reauthorization of funding for QECBs.

2.4.5 Qualified School Construction Bonds

Qualified School Construction Bonds (QSCBs) were created in 2009 under the American Recovery and Reinvestment Act (ARRA). [19] As is the case for the other bonds discussed in this section, QSCBs originally were designed as tax credit bonds. The proceeds from a QSCB can be used for school construction, rehabilitation, and repair, as well as land acquisition to site a school. Energy efficiency and renewable energy projects are permissible uses of bond proceeds under this definition. The first QSCB issued in the United States was from the San Diego Unified School District in 2009 (for $39 million) in combination with Capital Appreciation Bonds.[6] Since then, a number of California school districts have issued QSCBs, including West Contra Costa County Unified School District, which issued a $25-million bond on June 10, 2010. [20]

As noted, with the passage of the 2010 HIRE Act, QSCBs can now be issued as taxable bonds with the issuer receiving a subsidy from the U.S. Treasury. As a result of this change, QSCB issuances have increased tremendously. In the first half of 2010, 167 QSCBs were issued for more than $2.5 billion. [21] This compares to three bonds for a total of $106 million in the first half of 2009 and a total of $2.8 billion for 2009.

In 2010, California received a QSCB allocation of $720 million and Oakland Unified School District received its own allocation of $24 million. [22] The application for local school districts to tap into this 2010 QSCB allocation was posted on the California Department of Education (CDoE) on October 1, 2010. The CDoE has reported that the program is oversubscribed; it now prioritizes the awards based on criteria established in the enabling legislation. Existing voter-approved bond authority is required to be eligible. Additionally, large school districts—such as Oakland Unified, which received a direct allocation from the IRS—are not eligible to apply. For more information, consult the California Department of Education website at http://www.cde.ca.gov/ls/fa/qs/2010qscboverview.asp.

In July 2010, a $12-million, 10-year QSCB bond with an interest rate of approximately 5% was issued by the California School Finance Authority on behalf of a San Diego, California, charter school, High Tech High. [23] The direct-pay federal subsidy is greater than 5%; therefore, the entire interest rate is offset, thus creating a true interest-free bond for the school. Note that, as of March 2011, there was no reauthorization of funding for QSCBs.

2.4.6 Qualified Zone Academy Bonds

Although similar to QSCBs in structure, Qualified Zone Academy Bonds predate the other tax credit bond programs and were created in 1997.[7] The QZABs are directed at schools serving significant numbers of low-income families. Qualified Zone Academy Bonds provide a source of funding that can be used for renovating school buildings, purchasing equipment, developing curricula, and training school personnel, but not for new construction.[8] There is an additional requirement of partnering with the private sector, which includes financial contributions. It is conceivable that QZABs could be used much like QSCBs to finance energy efficiency and solar projects. The additional requirements for QZABs, however, could make them a less flexible instrument than a QSCB.

---

[6] Goldman Sachs, "Overview of Tax Credit Bonds" (May 2009). http://www.nast.net/2009TreasuryMgmt/Files/WED%20MarvinMarkus.pdf. Accessed January 15, 2011.

[7] Taxpayer Relief Act of 1997, section 226(a). Available at http://frwebgate.access.gpo.gov/cgi-bin/getdoc.cgi?dbname=105_cong_public_laws&docid=f:publ34.105. Accessed June 9, 2010.

[8] U.S. Department of Education. Qualified Zone Academy Bond. http://www2.ed.gov/programs/qualifiedzone/index.html.

In 2010, the California allocation of QZABs was $163 million. [24] Individual school districts can apply to the state for an allocation, or districts can jointly apply. [25] Note that, as of March 2011, there was no reauthorization of funding for QZABs. For more information, see the California Department of Education website at http://www.cde.ca.gov/ls/fa/qz/introd.asp.

3 Third-Party Financing

The use of third-party financing to install large PV systems is common in California, including by K–12 public school districts. For example, in August 2010, the San Diego Unified School District board approved the use of third-party financing to install 5.2 MW of solar on more than 80 school district rooftops. [26] This adds to the 4 MW of PV that the district has already installed. [26] Third-party financing is particularly useful in helping non-tax-paying entities, such as school districts, implement solar projects that cannot otherwise benefit from federal incentives. Using solar power purchase agreements (PPAs) and, possibly, energy savings performance contracting, districts can host on-site PV systems without any up-front capital investment.

3.1 Power Purchase Agreement

Under the terms of a solar PPA, the solar developer/investor owns, operates, and maintains the PV system and sells 100% of the solar electricity produced to the host (school district) at a fixed price for a negotiated term of up to 20 years. The federal tax incentives available to businesses—the business energy investment tax credit (ITC) and accelerated depreciation—can offset 50% or more of the installed cost of a PV system. [27] The PPA provider can then pass a portion of the savings on to the school in the form of a lower PPA cost of electricity. As a result, the third-party ownership model can be a cost-effective arrangement for many public entities that are interested in pursuing solar but lack access to the necessary funding or prefer to forego ownership for other reasons. Additionally, buyout options can be negotiated into the contract for the host to purchase the system sometime after 6 years and up through the end of the PPA term at the PV system's fair market value.

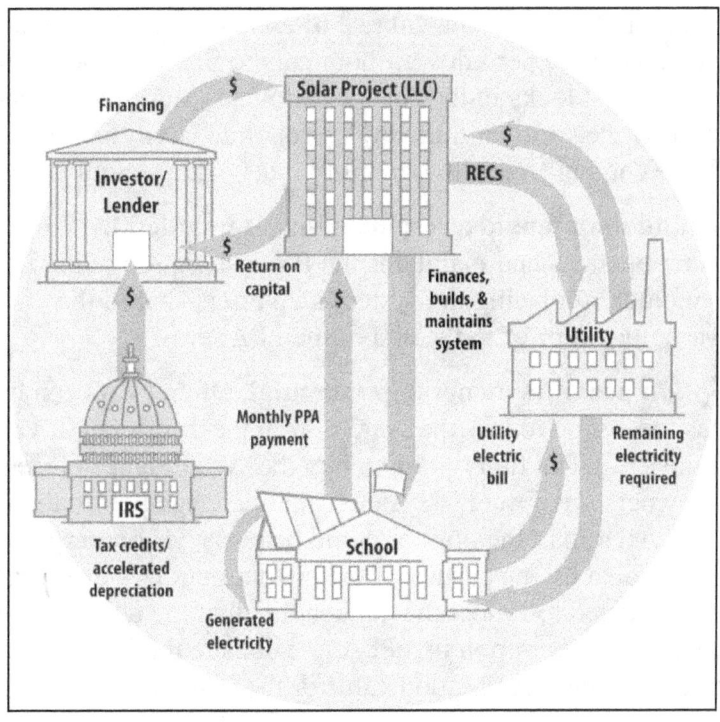

Figure 1. PPA flowchart (NREL 2010)

3.1.1 Advantages of the Third-Party Power Purchase Agreement Model for Solar

There are both advantages and disadvantages associated with third-party ownership models and solar PPAs. [28] Some of the commonly recognized benefits include the following:

- **Ability to benefit from the federal Business Investment Tax Credit.** As noted, commercial entities can benefit from the 30% ITC. By lowering the cost of the project to the solar developer and its investors, a lower PPA price can be offered to the public-sector host of the PV system.

- **Ability to benefit from modified accelerated cost recovery system (MACRS).** Photovoltaic installations can be depreciated over a 5-year period rather than over the expected useful life, which is much longer. Depreciation is treated as an expense for accounting purposes and reduces the income that is subject to taxes. As it relates to PV projects, the impact of depreciation usually is greater losses for the investors, which then are used to offset other taxable gains. Like the ITC, the host benefits from accelerated depreciation in that it could allow for a lower price per kilowatt-hour of electricity in the PPA.

- **No up-front capital investments.** Although installed costs are declining, the required initial investment to install a PV system is still significant, even after rebates. The cost of a 100-kW PV system on a middle school, for example, can exceed $500,000. Using the third-party PPA model, it is the solar developer and investors that finance and own the system, thus eliminating the need for the host to invest its own capital into the project.

- **Stable and predictable electricity prices for 20 years.** Power purchase agreements are commonly structured with an initial price per kilowatt-hour of electricity in the

17

first year, combined with an annual rate of escalation in the range of 2% to 5%. Alternatively, the price per kilowatt-hour can be fixed for the entire term of the PPA. Regardless, the host locks in the cost per kilowatt-hour of solar-generated electricity for the length of the contract. In today's economic environment, the initial PPA price must likely be competitive with the utility rates that a school is currently paying.

- **Operation and maintenance responsibility is handled by the system owner.** The system owner operates and maintains the PV system, removing this burden from the host. This includes replacing the system's inverters should they fail after the standard 10-year warranty but prior to the end of the PPA term.

- **Buyout option provides ownership potential.** Often PPAs can be structured so that the host has the option to buy the system from the developer at various points during the life of the PPA. The first option to buy the system takes place sometime after year 6, because ownership of the PV system cannot change before then without significant tax penalties. After that, the options could be every year, every 5 years, or whatever period is negotiated by the parties. If the buyout option is exercised, then the price should be discounted to reflect the tax benefits that the developer has received during the first 5 years. It is common in a PPA to calculate the buyout price as the greater of either a predetermined termination value or the system's fair market value.

- **Risk avoidance.** The risk of electricity production is borne by the PPA provider. The host only is obligated to purchase what the system produces. Additionally, the PPA provider commonly guarantees a certain level of minimum production of electricity, compensating the host for any shortfall. This is especially important if retail electricity rates are greater than the PPA rates, as the host would have to purchase more expensive power from the utility to make up the shortfall of the PV system.

3.1.2 Disadvantages of Third-Party Ownership

- **No free electricity.** Although the PPA price will ideally be less than retail utility prices, the host does not own the PV system; therefore, it will continue to pay for all of the electricity consumed at the facility.[9] This stands in stark contrast to owning a PV system, which generates "free electricity" (finance costs notwithstanding). In the case of a school district, which has access to funds that it doesn't have to repay directly (e.g., taxpayer-financed bonds, transfers from the state), owning a PV system reduces utility bills and frees up cash in the general fund to be used for other purposes.

- **No ownership of the "clean" energy attributes produced by a PV system.** Whoever owns the system claims its environmental benefits, unless those benefits have been sold to another party such as the utility. If a school district has signed a PPA, it cannot make explicit environmental claims such as being "solar-powered" unless the PPA allows the district to retain the renewable energy certificates. Allowing the district to retain the RECs, however, often can make a transaction unattractive for the solar developer. Therefore, electricity-only PPAs are most common. If the solar RECs have not been bundled with the electricity, public claims

---

[9] A district is obligated to purchase all the electricity produced by the PV system it hosts. If additional electricity is required, then it must be purchased from the local utility at the utility's standard rates.

of being solar-powered must be tempered, given that only the owners of the RECs can make such a claim. One solution is to purchase "replacement RECs"—usually cheaper wind or biomass RECs—to "green up" the project.

- **Transaction costs are high.** Negotiating a PPA is very labor intensive. An RFP is developed and issued to select a solar developer. The PPA and the lease agreement must then be negotiated with the winning bidder. This negotiation process easily can take 6 months or more. To recoup some of these transaction costs, some PPAs include a requirement that the solar developer must reimburse the host for expenses incurred. These costs, of course, are in turn recouped by the developer in the form of an increased PPA price. However, this could be a way to develop internal support for a transaction.

- **Project will likely need a large, anchor PV system.** The PPA providers will seek the opportunity to install one or more large PV systems in a school district for the transaction to benefit from economies of scale. Placing numerous small PV systems on many school buildings is unlikely to be cost effective. Ideally, for example, a high school or maintenance facility that can host a system as large as 1 MW to anchor a system-wide PPA project could be required. In the absence of a large installation, costs will increase. Projects that rely on a number of small systems also risk falling apart should the "anchor" drop out.

- **Facility access by third parties is necessary.** The developer and its subcontractors need access to the site to install the PV system and then to maintain it over time. For school districts this often must be coordinated so that students and faculty are not disrupted during the installation process. For certain facilities, this might be a concern; for others, such as a bus maintenance facility, it could be less so.

In cases where a public entity has signed a PPA, it is because the advantages outweigh the disadvantages. Alternatively, the lack of funding makes a third-party financed transaction the only realistic solution. If funding is obtained in the future, then ownership can be acquired by exercising the buyout option.

3.1.3 California Case Studies of Third-Party Financing Solar on K-Through-12 School Districts

In addition to the information contained in the following case studies, this document contains copies of the signed PPAs: https://www.musd.org/cms/page_view?d=x&piid=&vpid=1217983977356.

3.1.3.1 San José Unified School District, San José, California[10]

In 2007, Chevron Energy Solutions entered into a partnership with the San José Unified School District (SJUSD) to install solar panels on school buildings. The genesis of the project was the initiative of a local high school in the district that was interested in installing PV. It then became

a district-wide effort. The SJUSD had the following goals for the project:

---

[10] Information for this section was obtained from the Chevron Energy Solutions website, http://www.chevronenergy.com/case_studies/sjusd.asp; a SJUSD press release, http://www.naesco.org/resources/casestudies/documents/SJUSD-Solar-Press%20Release-final.pdf; and an interview with a representative of the school district (January 1, 2010) (on file with author).

- Deliver general fund savings

- Create education opportunities

- Demonstrate environmental stewardship and leadership.

In partnership with BankAmerica, the institution that financed and owns the PV installations, Chevron is installing a total of 5.5 MW of solar at 14 different sites across the district in three phases. Four high schools will host a total of 2 MW, and the remaining 10 sites will host 3.5 MW. Many of the sites are shade structures on parking lots, and the others are rooftop installations. BankAmerica is capturing the tax benefits as well as $11 million in incentives from the CSI Program. Chevron Energy Solutions is under contract to operate, monitor, and maintain the installations during the life of the PPA.

Solar energy is being incorporated into the district's science curriculum, and each of the 14 sites will have an educational display that includes system monitoring and real-time production information. The district expects to reduce energy costs by 30% during the life of the transaction (25 years) and save $25 million. Additionally, 100,000 metric tons of carbon dioxide will be avoided. Key design elements of the program are listed below:

- The district signed the PPA with the solar developer and is the party responsible for purchasing the solar electricity.

- The district negotiated an easement at each of the schools stipulating the conditions for third-party access and operation.

- From initial discussions to the first installation, the process took 18 months.

- Significant coordination was necessary with the selected schools during the pre-construction and construction phases because the installations took place during the school year.

- Several neighbors near one of the schools expressed concerns about the aesthetics of the solar installations. After viewing a computer-generated rendition, however, the neighbors ultimately supported the project.

- Some schools wanted to host PV systems but could not participate because they could site only small systems and not the large-scale capacity required for economies of scale to make the project "pencil out" for the investor.

- Initially, there was a great deal of skepticism on the part of the onsite building maintenance staff that had to be overcome. Installed systems have been relatively hassle free, however, so the project is currently meeting expectations.

- The maximum amount each system generates as a percentage of the building's electricity load is roughly 30% to 40%. The district has a net-metering agreement with the local utility.

- The district might be interested in buying the systems outright before the end of the contract, possibly using bond financing.

- The school district contracted with a third party to conduct independent inspections of the systems after they were installed.

3.1.3.2 Milpitas Unified School District, Milpitas, California[11]

In 2007, Milpitas Unified School District (MUSD) began discussions with Chevron Energy Solutions to carry out energy efficiency investments and install PV systems on school buildings. The district had the following four key objectives:

- Demonstrate economic leadership (general fund savings)
- Demonstrate environmental stewardship
- Create educational opportunities
- Receive positive public recognition and perform community outreach.

Figure 2. Solar PV array hosted by Milpitas Unified School District. *Photo by John Cimino*

The project consists of 3.4 MW of PV installations at 14 of the district's sites. These systems will meet 75% of the school district's annual electricity needs and 100% of its peak electricity needs during the summer. The installations are designed as both shade and carport parking structures. As with San José Unified, each site has an educational display showing system performance. Additionally, solar energy is integrated into the fifth-grade and sixth-grade curriculum. [29] BankAmerica financed and owns the PV installations and receives the tax benefits. The bank also received $4.2 million in CSI incentives.

The MUSD estimates that the system will save the district $12 million over the life of the project by reducing annual energy costs by 22%. The project will also reduce carbon dioxide emissions by 23,600 metric tons. The PV systems are assisting the school district in meeting California's Grid Neutral Initiative. [30] A phone interview was conducted with Director of Maintenance Operations and Transportation for MUSD, John Cimino, as part of a similar solar for schools report and revealed that the PV systems are producing more energy than guaranteed in the

---

[11] Information for this section was obtained from the Chevron Energy Solutions website, http://www.chevronenergy.com/case_studies/musd.asp.

contract, resulting in additional savings to the district. According to Mr. Cimino, the project has been a win-win for all parties involved, and was a fiscally responsible venture for the district as well as an environmental-stewardship measure.

3.1.4 Third-Party Power Purchase Agreements and New Market Tax Credits

The New Market Tax Credit (NMTC) is a mechanism by which private capital is channeled into low-income neighborhoods with the express intent of promoting economic development and jobs. [31] An investor in a community development entity (CDE) will benefit from a 39% federal tax credit over 7 years, in addition to the actual returns on the investment itself. The CDE, in turn, uses this investment to make either equity investments or loans to qualified projects within qualified neighborhoods. Although not a traditional source of capital for solar projects, certain public-sector projects are partnering with CDEs to finance PV installations, including the City of Denver, [32] Denver Public Schools, and Salt Lake County, Utah.

In the Denver case, a solar developer was able to obtain low-cost loans from a local CDE to finance a portion of what will be 3.9 MW of solar installations on city buildings and schools. The low-interest loans from the CDE reduced the cost of electricity in the power purchase agreement by 5% to 15%, depending on the project. More information on the City of Denver's NMTC project can be found at http://www.nrel.gov/docs/fy10osti/49056.pdf.

3.2 Energy Savings Performance Contracting

In 2008, $2.8 billion, or 69% of the total revenue for the energy savings performance contract industry, was generated by projects with municipal and state governments, universities and colleges, K–12 schools, and hospitals. [33] This illustrates that ESPCs are a viable mechanism to fund energy efficiency investments for public entities. The more difficult question is their applicability to solar energy projects.

An ESPC is a contract between a building owner (e.g., a school district) and an energy service company (ESCO) to carry out energy efficiency (including renewable energy) investments. The ESCO conducts a comprehensive energy audit for buildings throughout the district and identifies improvements to save energy. [34] In consultation with the schools, the ESCO designs and constructs projects that meet the district's needs. The ESCO guarantees that the improvements will generate energy cost savings sufficient to pay for the project over the term of the contract. [34] After the contract ends, all additional cost savings accrue to the district. [34] The energy service company can either finance the project or partner with a third party to finance it. Alternatively, the school district itself can finance the project and repay the debt with the guaranteed savings from the performance contract.

According to the National Association of Energy Service Companies (NAESCO), energy service companies handle the following tasks.

- Develop, design, and arrange financing for energy efficiency projects
- Install and maintain the energy efficient equipment involved
- Measure, monitor, and verify the project's energy savings
- Assume the risk that the project will save the amount of energy guaranteed.

These services are bundled into the project's cost and are repaid through the dollar savings generated. [35]

3.2.1 Incorporating Photovoltaics into an Energy Services Performance Contract

There are various approaches to including photovoltaics in ESPCs. Solar projects are usually only feasible within an ESPC with the help of incentives, rebates, or other forms of capital that can contribute to reducing the amount of financing required for the project. The ESCO can be a valuable resource to identify these grants, rebates, and incentives. One benefit of the ESPC model is the ability to bundle many energy efficiency measures from several buildings across a school district into one large performance contract. This method can leverage savings to reduce the payback period of a solar system that if implemented as a stand-alone project would not be feasible. This is possible because ESPCs use an average of the payback of all conservation measures included to determine the contract term. This is the most common method to implement small-scale solar projects in ESPCs.

3.2.2 Including Photovoltaics in an Energy Services Performance Contract

The Roslyn School District in New York has partnered with an ESCO in a performance contract that will save $230,000 annually over 15 years and capture $130,000 in solar and lighting state rebates. [36] Improvements to the schools across the district include building envelope and insulation improvements, lighting upgrades, boiler and heating system upgrades, and two 11-kW PV systems. [36, 37]

Although this anecdotal example illustrates that PV installations have been installed as part of an ESPC, in general it has proven to be difficult, especially for larger installations. One issue is that the return on investment for projects that include a PV system bundled with energy efficiency investments such as lighting; heating, ventilating, and air-conditioning (HVAC) controls; and chiller upgrades could still exceed the requirements of the project sponsors. A second issue is that title of the equipment installed under an ESPC normally transfers automatically to the public agency upon the completion of the work. This impacts the ESCO's ability to benefit from the federal tax credits because the intended owner of these assets is a tax-exempt, public entity. To work around this issue, an ESPC could be structured whereby the ESCO immediately transfers title to all of the energy efficiency equipment, but retains ownership of the PV system for at least 6 years to allow for the tax benefits to vest. After the 6-year term, the ESCO could sell the PV system to the school district at fair market value. Anything less than FMV could trigger the recapturing of tax benefits earned by the ESCO. Finally, although an ESCO might have expertise with a wide range of energy efficiency investments, it might be less familiar with solar projects, thus adding complexity to the transaction.

A possible alternative to the ESCO retaining title to the PV system for at least 6 years is to bundle the physical installations of both the PV systems and the energy efficiency projects, but separate the financing mechanisms into a performance contract and a PPA. The guaranteed savings under the ESPC would pay for the energy efficiency investments. In parallel, the building owner would purchase the electricity generated by the PV system under a PPA rather than buying the system outright. This preserves the shorter return-on-investment timeline for the energy efficiency improvements, avoids the need to purchase the system at fair market value at the time, and also allows the federal tax credits to be monetized through the PPA.

23

Despite these complexities, ESCOs have been expanding the types of technologies included in ESPCs. In 2006, 10% of the ESCO industry revenue came from onsite renewable energy projects. [33] By 2008, this had increased modestly to 14% of total revenues. [33] The individual renewable energy technologies themselves were not broken out in this particular study, but it does appear that the ESPC industry is increasing its expertise in this area.

3.3 Resources

The best resource for additional information on ESPCs is the Energy Services Coalition. [38] Its website provides a variety of ESCO template documents and is available at http://www.energyservicescoalition.org/resources/model/index.html. For additional information, please consult the National Association of Energy Service Companies (http://www.naesco.org/), the State Energy or Commercialization Office, and the Status of ESPC Enabling Legislation in the United States (http://www.ornl.gov/info/esco/ legislation/newesco.shtml).

Assistance from a national laboratory also can be accessed through the DOE Technical Assistance Program (http://www1.eere.energy.gov/wip/assistance.html). Federal ESPC best practices and guidance documents are valuable resources that often can be modified for local government initiatives and can be found on the DOE Office of Energy Efficiency and Renewable Energy (EERE) Federal Energy Management Program (FEMP) Resources website at http://www1.eere.energy.gov/femp/financing/ espcs_resources.html.

4 Summary

This report presents a number of energy efficiency and renewable energy options that are available to school districts as they implement their Solar Master Plans. Both direct-purchase and third-party finance alternatives are feasible, depending on the particular circumstances of each district. In certain cases the use of the various tax credit bonds will be limited to those districts with allocations in hand. Given its various allocations, Oakland Unified, for example, is well positioned to compare a variety of tax-credit bond options. Depending on available funding, third-party finance options could also be a course of action to pursue, even if eventual ownership in the medium term of the PV systems is the desired outcome.

5 References

1. *California Solar Initiative—Statewide Trigger Tracker.* Go Solar California. http://www.csi-trigger.com/. Accessed June 8, 2010.

2. *California Solar Initiative Program.* California Public Utilities Commission, June 2010. p. 27. http://www.gosolarcalifornia.ca.gov/documents/CSI_HANDBOOK.PDF. Accessed June 8, 2010.

3. Mortensen, J. *Factors Associated with Photovoltaic System Costs.* NREL/TP-620-29649. Golden, CO: National Renewable Energy Laboratory, June 2001, p. 3.

4. "A Guide to Installing a Solar Electric System." *Seattle City Light*, August 2009. p. 9. http://www.seattle.gov/light/conserve/cgen/docs/SCL_SolarGuide.pdf. Accessed June 8, 2011.

5. Christensen, C.; Barker, G. "Effects of Tilt and Azimuth on Annual Incident Solar Radiation for United States Locations." Presented at the 2001 Solar Energy Forum, Washington, DC.

6. *Glossary.* Database of State Incentives for Renewables and Efficiency. http://www.dsireusa.org/glossary/. Accessed June 8, 2011.

7. Kelly, T. Email communication, September 2, 2010.

8. *Energy Conservation Assistance Act.* http://www.energy.ca.gov/contracts/PON-10-601/PON-10-601_Notice_for_new_3pct.pdf. Accessed June 8, 2011.

9. *1603 Program—Payments for Specified Energy Property in Lieu of Tax Credits.* U.S. Department of the Treasury. http://www.treasury.gov/initiatives/recovery/Pages/1603.aspx. Accessed June 8, 2011.

10. *Planning Opportunity: Treasury Grant Guidance Permits Leasing to Governments and Tax-Exempts.* Hunton & Williams, January 2010. http://www.hunton.com/files/tbl_s10News/FileUpload44/16852/planning_opportunity_treasury_grant_guidance.pdf. Accessed June 8, 2011.

11. *Status of Funds.* California Department of General Services. http://www.documents.dgs.ca.gov/OPSC/Resources/Funds_Status.pdf. Accessed June 9, 2011.

12. *High Performance Incentive Program.* California Department of General Services. http://www.dgs.ca.gov/dsa/Programs/progSustainability/hps.aspx. Accessed June 9, 2011.

13. Schroeder, P. "Bill Comes to BABs Rescue." *Bond Buyer Online*, July 29, 2010. http://www.bondbuyer.com/issues/119_393/babs_bill_levin-1015368-1.html. Accessed June 15, 2011.

14. *Energy Policy Act of 2005.* http://frwebgate.access.gpo.gov/cgi-bin/getdoc.cgi?dbname=109_cong_bills&docid=f:h6enr.txt.pdf. Accessed June 9, 2011.

15. *First Known Use of QECBs Will Save Yolo County at Least $8.7 Million over the Next 25 Years.* Renewable Energy Project Finance. http://financere.nrel.gov/finance/content/first-known-use-qecbs-will-save-yolo-county-least-87-million-over-next-25-years. Accessed June 9, 2011.

16. IRS Notice 2009-29, "Qualified Energy Conservation Bond Allocations for 2009." http://www.irs.gov/irb/2009-16_irb/ar10.html. Accessed June 9, 2011.

17. California Debt Limit Allocation Committee (CDLAC) QECB Program. 2009 Presentation. http://www.treasurer.ca.gov/cdiac/seminars/20091008/6b.pdf. Accessed June 9, 2011.

18. California Debt Limit Allocation Committee. http://www.treasurer.ca.gov/cdlac/staff/20090722/7.pdf. Accessed January 15, 2011.

19. Qualified School Construction Bond Allocations for 2009. http://www.irs.gov/pub/irs-drop/n-09-35.pdf. Accessed June 9, 2011.

20. "Qualified School Construction Bonds. List of Issuers." *The Bond Buyer,* August 2010. http://www.bondbuyer.com/pdfs/QSCB.pdf. Accessed June 9, 2011.

21. Schroeder, P. "After Direct-Pay Option Is Added, QSCBs Stay Strong." *The Bond Buyer,* August 2010. http://www.bondbuyer.com/issues/119_400/-1015839-1.html. Accessed June 9, 2011.

22. *2010 Allocations to States of Volume Cap for Qualified School Construction Bonds.* http://www.treasury.gov/press/releases/reports/bonds.pdf. Accessed January 15, 2011.

23. "California Charter School Scores Interest-Free QSCB Deal." *High Beam Research*, July 30, 2010. http://www.highbeam.com/doc/1G1-232996621.html. Accessed June 15, 2011.

24. IRS Notice 2010-22, "Qualified Energy Conservation Bond Allocations for 2010." http://www.irs.gov/irb/2010-10_IRB/ar07.html. Accessed June 15, 2011.

25. *QZAB Allocations.* California Department of Education. http://www.cde.ca.gov/ls/fa/qz/introd.asp. Accessed June 9, 2011.

26. Magee, M. "Schools getting a solar jolt." *San Diego Union Tribune*, August 9, 2010.

27. Bolinger, M. *Financing Non-Residential Photovoltaic Projects: Options and Implications.* Lawrence Berkeley National Laboratory, January 2009. http://eetd.lbl.gov/ea/EMS/reports/lbnl-1410e.pdf. Accessed June 9, 2011.

28. Cory, K., Coughlin, J.; Coggeshall, C. *Solar Photovoltaic Financing: Deployment on Public Property by State and Local Governments.* NREL/TP-670-43115. May 2008. http://www.nrel.gov/docs/fy08osti/43115.pdf. Accessed June 9, 2011.

29. Cimino, J. "Milpitas Unified School District Sustainability Program." Presented at Milpitas Unified School District, 2009.

30. "Milpitas Unified School District Partners with Chevron and Bank of America Corporation on 3.4MW Solar and Energy Efficiency Program Expected to Save $12 Million for Education." *Business Wire*, June 2008. http://www.businesswire.com/portal/site/home/permalink/?ndmViewId=news_view&newsId=20080625005994&newsLang=en. Accessed June 9, 2011.

31. *New Market Tax Credits Program.* http://www.cdfifund.gov/what_we_do/programs_id.asp?programID=5. Accessed June 9, 2011.

32. *Financing Solar Installations with New Markets Tax Credits: Denver, Colorado.* NREL Energy Analysis Fact Sheet. September 2010. http://www.nrel.gov/docs/fy10osti/49056.pdf. Accessed June 9, 2011.

33. Satchwell, A.; Goldman, C.; Larsen, P.; Gilligan, D.; Singer, T. *A Survey of the U.S. ESCO Industry: Market Growth and Development from 2008 to 2011.* LBNL-3479E (June 2010): 11. http://eetd.lbl.gov/ea/emp/reports/lbnl-3479e.pdf. Accessed June 9, 2011.

34. *Energy Savings Performance Contracting (ESPC).* Federal Energy Management Program (FEMP). http://www1.eere.energy.gov/femp/financing/espcs.html. Accessed June 9, 2011.

35. *Resources.* National Association of Energy Service Companies (NAESCO). http://www.naesco.org/resources/esco.htm. Accessed June 9, 2011.

36. *A Major Investment in a Greener Future: Energy Performance Contract Will Save on Energy; No Net Cost to Taxpayers for Improvements.* Roslyn School District, March 2009. http://www.roslynschools.org/capital/epc.htm. Accessed June 9, 2011.

37. Mohrman, T. Assistant to the Superintendent for Operations at the Roslyn School District. Email correspondence. August 27, 2010.

38. *Energy Performance Contracting.* Energy Services Coalition. http://energyservicescoalition.org/index.html. Accessed June 9, 2011.

Appendices

Appendices for this report are contained in a separate document; some of these resources were purchased or obtained exclusively for the three school districts and cannot be made publicly available. However, some of the documents are available online, and in those cases, the websites are provided below. These references include a number of pertinent documents related to financing solar installations on schools and other public facilities. A list of these documents follows.

Request for Proposal for Procurement of Photovoltaic on Public Schools

- San Ramon Valley Unified School District: http://www.srvusd.net/solar

- Mount Diablo Unified School District

ESPC Documents

- RFP Template for ESPC:
  http://www.energyservicescoalition.org/resources/model/index.html

- Energy Performance Contract Template:
  http://www.energyservicescoalition.org/resources/model/index.html

- Financing Solicitation Template:
  http://www.energyservicescoalition.org/resources/model/index.html

Third-Party PPA Documents

- NREL Checklist: http://www.nrel.gov/docs/fy10osti/46668.pdf

- SolarTech PPA Template

- SolarTech Lease Template

- Milpitas Unified School District PPA with BankAmerica:
  https://www.musd.org/cms/page_view?d=x&piid=&vpid=1217983977356

New Markets Tax Credit

- NREL Fact Sheet on the City of Denver :
  http://www.nrel.gov/docs/fy10osti/49056.pdf

- NREL Fact Sheet on Yolo County: http://www.nrel.gov/docs/fy11osti/49450.pdf

Office of Public School Construction Funds

- School Facility Program requirements for the New Construction Grant:
  www.documents.dgs.ca.gov/opsc/Resources/SFP_NC_Rqmnts.pdf

- School Facility Program requirements for the Modernization Grant:
  www.documents.dgs.ca.gov/opsc/Resources/SFP_Mod_Rqmnts.pdf

- Checklist for High Performance Incentive (HPI) Projects:
  www.documents.dgs.ca.gov/dsa/other/HPI_Checklist_rev02-07-10.pdf

- DSA High Performance Incentive (HPI) Scorecard and Guidelines:
  http://www.documents.dgs.ca.gov/dsa/other/GL-5_HPI.pdf